AGS®

2002

HOW TO BUILD YOUR SELF-ESTEEM

Braham

Barbara J. Braham
Consultant/Professional Public Speaker
Columbus, Ohio

Merle Wood
Education Consultant
Formerly of Oakland Public Schools
Lafayette, California

AGS®
American Guidance Service, Inc.
Circle Pines, Minnesota 55014-1796
1-800-328-2560

Printed in the United States of America

ISBN 0-7854-1343-X

Product Number 50150

A 0 9 8 7 6 5 4 3 2

This book is printed on recycled, acid-free paper that meets Environmental Protection Agency Standards.

ACKNOWLEDGMENTS

Cover Photo/Illustration: [1]Westlight/© Walter Hodges
[2]Westlight/© Warren Morgan

PREFACE

There is nothing more crucial to a person's success than high self-esteem. Yet how many people can honestly look themselves in the mirror and say, "I like myself?" HOW TO BUILD YOUR SELF-ESTEEM takes students through a step-by-step process that leads to greater self-confidence, increased feelings of self-worth and enhanced self-respect. The text-workbook is written specifically for the adult learner and is designed to permit self-paced, individualized instruction and foster student success.

The major premise of HOW TO BUILD YOUR SELF-ESTEEM is that your level of self-esteem can be increased. The book helps students see the critical difference between self-esteem and self-image. When they can break out of the self-image trap, they can stop worrying about what other people think.

The book outlines how most people engage in negative self-talk, and presents students with the tools to change negative self-talk into positive self-talk. The last unit stresses the commitment and discipline necessary to maintain high self-esteem.

HOW TO BUILD YOUR SELF-ESTEEM is highly practical with proven techniques for helping students change how they feel about themselves. The mastery of these concepts will increase their success in all other academic courses.

SPECIAL FEATURES

HOW TO BUILD YOUR SELF-ESTEEM is designed specifically to help you invest in the future of your adult learners and to meet your instructional needs. Some features of the text-workbook include the following:

- A larger typeface is used to make the text-workbook easier for students to use and to read. Pages are colorful and uncrowded.
- Competency-based methodology is used. Clear objectives are presented followed by short segments of instruction. These are followed by student activities for immediate reinforcement.
- Content and examples relate to adult-level, real-life issues and skills.
- Pretests and post-tests, with answers, are included for self-evaluation.
- Study breaks are included to provide refreshing and useful information that contributes to the general literacy of the student.

- Abundant exercises are included, each designed so that students experience frequent and meaningful success.
- Goals are listed for each exercise to provide motivation and direction.
- All exercises are supported with Bonus Exercises for students who need a second chance to succeed.
- Answers to all exercises are included to facilitate independent, self-paced learning.
- Personal progress is recorded by the student after completing each exercise.
- Individual success is measured by evaluation guides in the student's Personal Progress Record.

INSTRUCTOR'S MANUAL

The Instructor's Manual provides instructional strategies and specific teaching suggestions for HOW TO BUILD YOUR SELF-ESTEEM along with supplementary bonus exercises and answers, additional testing materials, and a certificate of completion.

Bonus Exercises. A bonus exercise, matching each exercise in the text-workbook, is provided in the manual. These bonus exercises make it possible for students to have a second chance to reach the goals set for each exercise. Answers to the bonus exercises are also provided in the manual. These materials may be reproduced for classroom use.

Testing Materials. Two additional tests are provided in the manual to allow for more flexible instruction and evaluation.

Certificate of Completion. Upon completion of HOW TO BUILD YOUR SELF ESTEEM, a student's success may be recognized through a certificate of completion. This certificate lists the skills and topics covered in this text-workbook. A certificate master is included in the manual.

CONTENTS

GETTING ACQUAINTED

You want to be happy. You want to feel successful. You want to enjoy your life. These things are all possible when you like yourself - when you have high self-esteem. HOW TO BUILD YOUR SELF-ESTEEM will show you how you can feel better about yourself.

Some people worry about what other people think, and feel bad about themselves. Some people talk to themselves in a negative, judgmental way. You're going to learn how to believe in yourself, use positive self-talk, and keep your self-esteem high in most situations.

HOW TO BUILD YOUR SELF-ESTEEM will show you the steps to take so that you can look yourself in the mirror and confidently say, "I like myself!"

HOW YOU WILL LEARN

HOW TO BUILD YOUR SELF-ESTEEM is written with you in mind. You will learn what you can do to feel better about yourself tomorrow, no matter how you feel about yourself today. There is a system used in this book that will help you learn.

Learn at Your Own Pace

You will progress through the lessons in this book at your own pace. You may move ahead faster, or go slower, than other students. But, don't be concerned about this. You are to work at *your* best speed.

Learn Skills Successfully

You are given objectives and goals for each unit. You will know what you are to accomplish. You will study a topic. Then you will complete an exercise. This lets you drill over what you have just learned. When you have shown that you know the topic, you will move on to the next topic. You will always know just how well you are doing as you move through each step in this book.

Complete Bonus Exercises

You may not reach your goal on every practice exercise. When this happens, review the lesson and then complete a Bonus Exercise. These exercises cover the same lessons as the practice exercises in this book. They give you a second chance to reach your goal. When you score higher on a Bonus Exercise than you did on the original activity, you may change your score on your

Personal Progress Record. Your instructor has copies of these Bonus Exercises and the answers to them.

Check Your Own Success

You will keep track of your own success. You will check all of your own work. The answers are in the back of this book. The colored pages make them easy to find. Always do the exercises *before* you look at the answers. Use the answers as a tool to verify your work—not as a means of filling in the blanks. You will record your scores on your own Personal Progress Record, which is also in the back of this book.

WHAT YOU WILL LEARN

As you study HOW TO BUILD YOUR SELF-ESTEEM, you will learn how to change negative thinking into positive thinking. When you can do this everyday, you will have high self-esteem. To learn this skill you will study these subjects:

Unit 1, What Is Self-Esteem?
Check your present level of self-esteem
How to stop worrying about what other people will think
3 ingredients of high self-esteem

Unit 2, Negative Thinking
How self-talk can raise or lower your self-esteem
3 types of shoulds
5 types of negative thoughts - and what to do about them

Unit 3, Positive Self-Talk
Meet your Fan Club
4 steps to powerful affirmations
How to talk back to your inner Judge

Unit 4, I Like Myself!
4 tips to maintain high self-esteem
3 ways to build the self-esteem of others
How to set SMART goals

SPECIAL FEATURES

HOW TO BUILD YOUR SELF-ESTEEM has a number of special features. These features will help you learn and apply the material successfully.

Checking What You Know

You can check what you already know about HOW TO BUILD YOUR SELF-ESTEEM before you start studying this book.

Checking What You Know lets you know what skills you need to improve upon. Then, when you complete this book, you will do an exercise called *Checking What You Have Learned*. By comparing these two scores you will see how much you have gained through your study.

Lift Your Spirits

This feature of the book suggests exercises you can do by yourself, or with others, to build your self-esteem and lift your spirits. It will help you put into practice what you are learning.

Putting It Together

The exercises in each unit are called Checkpoints. These exercises will help you check your understanding of a specific topic before continuing. At the end of each unit you will find a section titled *Putting It Together*. This section contains several exercises that are similar to the Checkpoints. They will help you reinforce the skills you learned in each unit.

Personal Progress Record

You will keep track of your own progress. Once you check your answers, you will record your score on your *Personal Progress Record* at the end of this book. After you finish a unit, you will be able to see your level of success.

Completion Certificate

When you finish your study in this book you may be eligible for a certificate of completion. Your instructor will explain to you the skill level required for this award.

READY TO START

You are now ready to start building your self-esteem. Through your study and completion of the exercises, you will learn to genuinely like yourself.

Your new skill will be of benefit to you both at home and at work. As you learn to accept yourself, you will also accept others. That means you'll have a better relationship with yourself, and with others.

Turn to page ix and complete *Checking What You Know*. Check your answers with the answers on page 47. Then begin *Unit 1, What is Self-Esteem*?

CHECKING WHAT YOU KNOW

Take this pre-test before starting HOW TO BUILD YOUR SELF-ESTEEM. The 25 questions will tell you how much you already know about self-esteem. They will also tell you what you need to learn.
There is no time limit, so take your time. When you finish, check your answers. Give yourself 2 points for each correct answer. Record your score on your Personal Progress Record. After finishing the book, you will be able to see how much you learned.

DIRECTIONS: Read each of the following statements. If a statement is true write a *T* in the space provided. If it is false write an *F* in the space provided.

1. ____ Self-esteem is how you feel about yourself.
2. ____ Being around someone with high self-esteem makes you feel good about yourself.
3. ____ People may have high, middle or low self-esteem.
4. ____ Self-image is when you try to imitate people.
5. ____ People with high self-esteem feel self-confident.
6. ____ You can have a positive self-image and low self-esteem.
7. ____ Most of your self-talk is positive.
8. ____ One way to motivate yourself is to say, "I should."
9. ____ "You ought to study more" is an example of a should statement.
10. ____ The more you use shoulds the better you will feel about yourself.
11. ____ Overgeneralizations are one type of negative thinking.
12. ____ When you call someone names, it is called fortune telling.
13. ____ Catastrophizing builds self-esteem.
14. ____ You have an inner Fan Club who builds your self-esteem.
15. ____ An affirmation helps you keep a positive focus.
16. ____ For an affirmation to work, you need to be able to visualize it.
17. ____ "I'm not scared," is an example of an affirmation.
18. ____ Affirmations are stated in the future tense.
19. ____ "I do good work," is an example of an affirmation.
20. ____ Affirmations are like self-esteem food.
21. ____ You can build your self-esteem by being around people who criticize you.
22. ____ Judging the past will build your self-esteem.
23. ____ To build the self-esteem of others, notice when they do something right.
24. ____ A SMART goal is tough.
25. ____ "Be a better parent" is an example of a SMART goal.

☞ *Check your work on page 47. Record your score on page 51.*

UNIT 1

What Is Self-Esteem?

WHAT YOU WILL LEARN

When you finish this unit, you will be able to:

- Define self-esteem.
- Define self-image.

Do you like yourself? Do you believe you can do anything you set your mind to do? Do you accept other people as they are?

These are all questions about your level of self-esteem. Could you answer "yes" to each question? If not, this book is for you! You can learn to like yourself. You can change your beliefs. You can learn to accept yourself and others.

When you can answer "yes" to each question, you will know you have high self-esteem.

SELF-ESTEEM: A DEFINITION

Self-esteem is how you feel about yourself. When you look at yourself in the mirror in the morning can you say, "I like myself!" Or, do you look in the mirror and think "I'm ugly," or "I wish I was different," or "I don't like myself."

If you don't like yourself very much, you have low self-esteem. If you like yourself most of the time, you have high self-esteem. Most people have self-esteem that is somewhere in the middle. Self-esteem goes up and down like a yo-yo. Some days you wake up, feel great, and the whole day goes great. Other days you start the day feeling bad, and you feel that way all day.

Check Your Self-Esteem

Read through each of the following statements. Put a *T* in the space provided if a statement is true for you. Leave the space blank if it is not true.

1. ____ I can admit a mistake.
2. ____ I can talk to people I don't know.

3. ____ I do what I believe is right even if others don't approve.

4. ____ I can accept a compliment and not feel uncomfortable.

5. ____ I can be myself around other people.

6. ____ I accept myself with all my faults and weaknesses.

7. ____ I can tell you my strengths.

8. ____ I can feel happy for someone else when he or she succeeds.

9. ____ I do not compare myself with others.

10. ____ I have peace of mind.

11. ____ I believe I am unique.

12. ____ I accept differences in others without judging them.

13. ____ I compliment others.

14. ____ I can say "I love you" to all the people I love.

15. ____ I love myself.

16. ____ I am comfortable being alone.

Count the number of statements you marked *True*. If you have 11 or more trues, you have high self-esteem. Answering true to 6-10 statements puts you in middle self-esteem. You have low self-esteem if you had 5 or fewer trues.

However you scored on the checklist you can learn to feel better about yourself. You can have more good days than bad days. You can feel like the people in Illustration 1-1 do.

You can usually tell how a person feels about him/herself. You can tell by how you react to the person. If you think to yourself, "Wow! I can't wait to see you again," you were with someone with high self-esteem. Being with him or her makes you feel good about yourself. But, if you think, "Boy, I hope I don't see him/her again soon," you were with someone with low self-esteem. If you didn't feel strongly one way or the other, the person probably had middle self-esteem.

If you have low self-esteem, you are often negative. You don't expect things to work out, and you may even stop trying. You don't feel in control of your life. You react to the things that happen to you with blame and feelings of worthlessness.

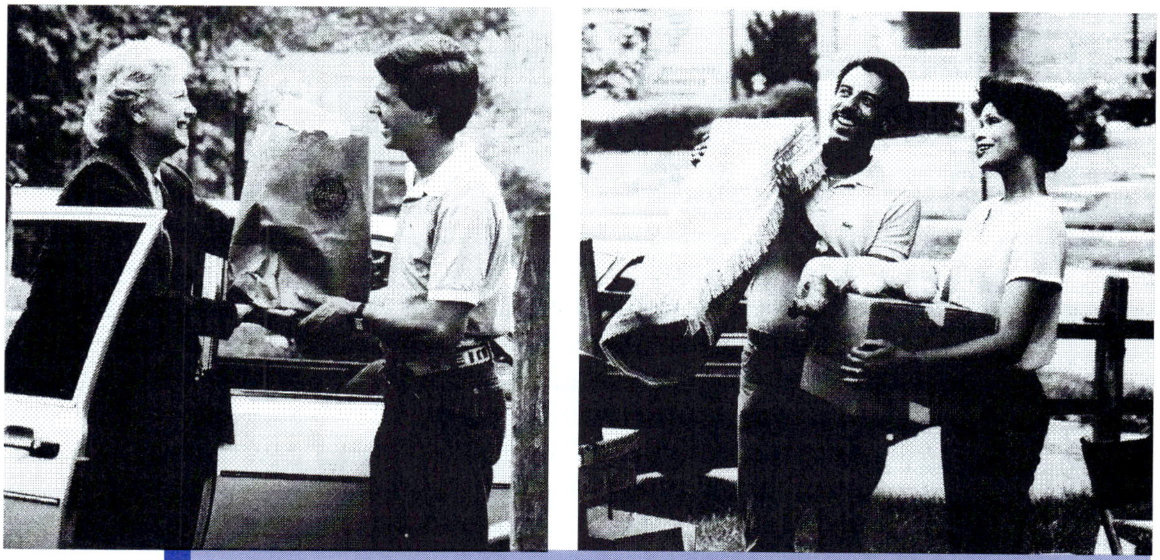

Illustration 1-1

Feeling Good Shows

People in the middle need the approval of others. They feel self-doubt. They hope that the doubt will go away if they can get others to like them. They listen more to others than to themselves. They are people pleasers.

If you have high self-esteem, you aren't afraid to be yourself. You can risk being different from the crowd. You can say what you think and feel. You aren't worried about what other people will think. Most of the time you feel self-confident. You have a positive attitude about life and feel in charge of your life.

CHECKPOINT 1-1

YOUR GOAL: Get 7 or more answers correct.

Read the list of words. Decide if they describe a person with high, middle, or low self-esteem. Write *High*, *Middle*, or *Low* in the space provided. An example has been completed for you. Give yourself one point for each correct answer.

- High risk-taker

1. ____________ trusting
2. ____________ suspicious
3. ____________ unsure
4. ____________ inferior
5. ____________ self-confident

6. ____________ self-doubt

7. ____________ self-hatred

8. ____________ I can

☞ *Check your work on page 47. Record your score on page 52.*

When you have high self-esteem, you decide that you are a valuable, worthwhile human being. Other people may like and approve of you, or they may not. You don't let them control how you feel about you, though. In other words, your feeling of self-esteem comes from inside of you. You can like you even if someone else doesn't! You can stop trying to please others. You can be yourself!

Self-Worth

Every person has worth. You have worth because you are alive. **Self-worth** is your belief that you are a valuable human being. You feel like you count. If your self-esteem drops, you can feel worthless. When you feel worthless, you think that you don't matter. You feel empty inside.

Some people increase their feelings of self-worth by doing things that they think are worthwhile. They do things that they think will make a difference. For example, if you work outside the home, you might get feelings of self-worth from your job. Tasks that you do at home like fixing the car, cleaning the house, raising children, or growing a garden can also give you self-worth.

Self-Respect

Self-respect is the opinion you have about yourself. If you behave according to what you believe is right, you can respect yourself. But if you believe one thing and act in a different way, you will lose respect for yourself. Loss of self-respect is the first step toward self-hatred.

LIFT YOUR SPIRITS

Think back to a time when you did something that you felt good about. You might have done something for another person. Or you might have achieved something. You could have made something. Remember everything you can about this time when you felt good about yourself. Then complete this sentence: I felt good about myself when ____________ because ____________. The next time you feel down on yourself, remind yourself of this time when you felt good.

Values are *your* beliefs about what is right. Different people have different values. Values help you decide what is important in life. For example, you might value honesty. As long as you tell the truth, you will have self-respect. But if you say you value honesty and then you tell lies, you won't respect yourself. If you do the same thing with other values, your self-esteem will suffer because you will lose your self-respect.

Self-Confidence

Self-confidence is believing in yourself. You have an attitude of "I can do it!" You trust yourself. Acting self-confident is difficult if you don't like yourself. If you feel worthless, you won't have self-confidence. One benefit of building your self-esteem is that you will feel more self-confident. When you meet people who are self-confident, you know they believe in themselves.

Self-worth and self-respect are both parts of self-esteem, as you can see in Illustration 1-2. You need self-worth and self-respect to feel good about yourself. When you like yourself, you act self-confident because you feel that way.

Illustration 1-2

Esteem Leads to Confidence

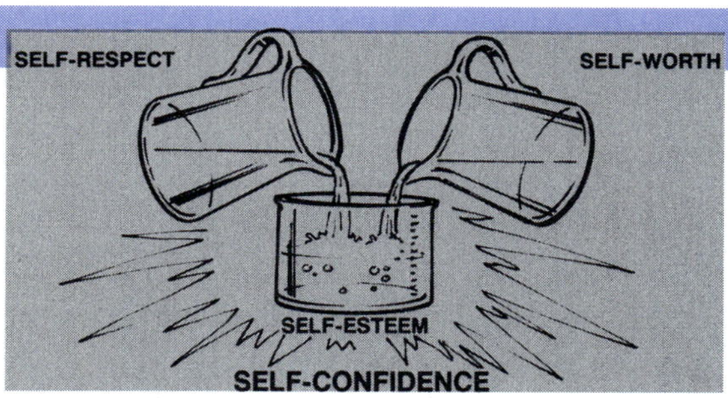

CHECKPOINT 1-2

YOUR GOAL: Get 4 or more answers correct.

Complete the following statements. Write the correct word or words in the space provided. Choose your answer from the list of words provided. An example has been completed for you. Give yourself one point for each correct answer.

self-esteem	self-worth
self-respect	values

- <u>Self-confidence</u> comes when you like yourself.

1. The set of beliefs that you use to tell right from wrong are your ____________.
2. If you act different than what you believe is right, you can lose your ____________.
3. You may get feelings of ____________ from what you do.

4. To have high self-esteem you need both ___________ and ___________.
5. People who appear to be self-confident also have high ___________.

☞ *Check your work on page 47. Record your score on page 52.*

SELF-IMAGE: A DEFINITION

Self-image is what you want other people to think of you. Self-image is often confused with self-esteem. The terms are different. Most people are very concerned about what other people think. They want other people to like them. They think that if other people like them, then they will like themselves.

Magazine, radio, and TV ads are designed to make you think that a positive image is the same thing as high self-esteem. Ads give the impression that if you wear a certain brand of clothes or drive a certain car you will be sexy or successful. Have you noticed that if you buy these products, how you *feel* about yourself doesn't change?

Self-image is not the real you. Self-image is an imitation. You try to imitate the people you admire. You try to make people believe you are the person you wish you were. For example, Pedro might be very sensitive and tender as a person. When he watches TV, he sees images of men who are tough and macho. If Pedro tries to be like the TV characters instead of himself, he will become concerned about his "image." He won't be himself.

Concern about his image hurts Pedro in two ways. First, he may not honor his values. To act tough he may do things that conflict with his sensitivity. If he behaves differently from his values, he can lose his self-respect.

Second, even if Pedro is successful creating the macho image, and everyone thinks he is tough, he won't feel good. Why not? Because no one knows the real Pedro. They like the image.

Self-image is a trap. To create a positive self-image is often to destroy your self-esteem. And, self-image makes you an addict. You become addicted to the opinions of other people. You look for them to tell you that you are "okay." You become addicted to what's "in." To be accepted, you change with each new fad. Liking yourself comes from the outside in. *If* others like you, *then* you can like yourself.

With self-image, whether you like yourself or not *depends*. Liking yourself depends on if you have the things the TV or radio says you need to be okay. It depends on what your friends and co-workers think about you. And liking yourself depends on what your family believes about you. You may never measure up.

If you have self-esteem, liking yourself comes from the inside out. Others like you because you like yourself.

If you *try* to have a positive self-image, you will try to impress other people. You will try to meet their expectations, hoping to be accepted. If you have high self-esteem, you will accept yourself and others. You will be true to yourself.

Some people try so hard to project a positive image that the opposite happens. They brag, seem selfish, act superior, and talk about themselves most of the time. You may feel like you want to get away from them. When you are around someone with high self-esteem, you feel good about yourself. You want to be around the high self-esteem person again and again.

LIFT YOUR SPIRITS

Napoleon Hill studied people who were successful in the 1920s and 1930s. He wrote a book about what he learned called Think and Grow Rich. Millions of people read his book and learned how to be successful. Napoleon Hill said, "If you think you can, or if you think you can't, you're right."

CHECKPOINT 1-3

YOUR GOAL: Get 8 or more answers correct.

Read each of the following statements. Decide if a statement describes self-esteem or self-image. Write the word *Esteem* or *Image* in the space provided. An example has been completed for you. Give yourself one point for each correct answer.

- Esteem — How you feel about yourself.

1. ________ I like myself.
2. ________ I can outdrink anyone.
3. ________ You make choices based on what you believe is right.
4. ________ I believe I can do what I decide I want to do.
5. ________ You try to impress other people.
6. ________ I accept myself.
7. ________ I compare myself to others.
8. ________ You worry about what other people will think.

9. __________ I have worth as a person.

10. __________ I try to be who other people want me to be.

☞ *Check your work on page 47. Record your score on page 52.*

Self-esteem comes from knowing and accepting yourself as you are. You stop trying to be someone you are not. When you don't accept yourself, you wish you were someone else. That is the image you try to project. You compare yourself with the person you wish you were, and you don't measure up. You feel there is something wrong with you. As long as who you are doesn't match with who you wish you were, you will feel emotional pain.

Where does the ideal person that you wish you were come from? From other people's expectations. Those people are usually parents, teachers, and others who were important to you when you were young. They told you how you *should* think, feel, and act. They did not always accept you as you were.

Your ideal self may also come from television or magazines. You may wish you were like a movie star, or a person you see pictured in a magazine.

When you have high self-esteem, you like who you are. You don't try to be someone else. You try to be the best that you can be.

LIFT YOUR SPIRITS

You can start today to change how you feel about yourself. How? By saying "I like myself" at least three times each morning and evening. Saying these words will feel strange at first. After a week or so, they will be easier to say. After a month, you will start to feel different about yourself. Try it!

WHAT YOU HAVE LEARNED

Self-esteem is how you feel about yourself. When you have self-worth and self-respect, your self-esteem is high. People with high self-esteem seem confident. Self-image is different from self-esteem. Your self-image comes from trying to please others. You try to meet others' expectations of you. When you *try* to have a good image, you lower your self-esteem.

PUTTING IT TOGETHER

ACTIVITY 1-1 YOUR GOAL: Get 8 or more answers correct.

Complete each of the following statements. Write the correct word or words in the space provided. Choose your answer from the list of words provided. An example has been completed for you. Give yourself one point for each correct answer.

depends	image	low	middle
self-esteem	self-image	self-worth	values

- People who like themselves have high self-esteem

1. If you don't take risks, see the negative side of things, and feel worthless, you have _______________ self-esteem.
2. If your self-esteem bounces like a yo-yo, you probably have _______________ self-esteem.
3. When you try to be someone you aren't, you are concerned about your _______________.
4. You can destroy your _______________ with your self-image.
5. With _______________, you think that if others like you, you will like yourself.
6. When you accept who you are, you have high _______________.
7. Achieving or accomplishing something adds to your feelings of _______________.
8. Most advertising wants you to project a positive _______________.
9. Your image _______________ on what other people think.
10. People with high self-esteem act according to their _______________, or beliefs, not according to what other people will think.

☞ ***Check your work on page 47. Record your score on page 52.***

ACTIVITY 1-2 YOUR GOAL: Get 8 or more answers correct.

Read each of the following statements. Choose the answer that is correct. Write the letter in the space provided. An example has been completed for you. Give yourself one point for each correct answer.

- __c__ Which one of these words is not related to self-esteem?
  a. self-confidence
  b. self-worth
  c. self-image

1. ______ Which word describes someone with high self-esteem?
   a. confident
   b. negative
   c. blaming

2. ______ Which words describe self-image?
   a. Bounces like a yo-yo
   b. Imitation — who you wish you were
   c. Accept yourself

3. ______ Which words describe how a person with a positive image may show low self-esteem?
   a. Brags, acts superior, talks a lot
   b. Confident, listens, trusts
   c. Smart, good job, lots of friends

4. ______ A person trapped by an image is addicted to:
   a. drugs
   b. approval
   c. TV

5. ______ What is self-respect?
   a. You act according to your values
   b. You can prove yourself to someone
   c. You achieve something

6. ______ Which of these activities would give you a feeling of self-worth?
   a. Reading a book
   b. Planting a garden
   c. a & b

7. ______ Which phrase does not describe self-image?
   a. The person you wish you were
   b. High self-esteem
   c. The person you think you should be

8. ______ Self-confidence
   a. Is knowing you are a valuable human being
   b. Comes from other people
   c. Is believing in yourself

☞ ***Check your work on page 47. Record your score on page 52.***

UNIT 2 Negative Thinking

WHAT YOU WILL LEARN

When you finish this unit, you will be able to:

- Change your shoulds to more positive self-talk.
- Change overgeneralizations to facts.
- Stop using labels to describe yourself.
- Replace fortune telling with goal setting.
- Identify examples of catastrophizing.

All day long you talk to yourself. You remind yourself to pick up milk at the store. You plan what you will do over the weekend. You rehearse what you will say to your boss. You remember things that happened in the past. These conversations that you have with yourself are called **self-talk**. Everyone does it.

Some people talk positively to themselves. Other people talk negatively to themselves. If your self-talk is negative, you will have low self-esteem. If your self-talk is positive, you will have high self-esteem.

In this unit you will learn how to tell if your self-talk is positive or negative. If it is negative, you will learn how to change it to positive self-talk.

SHOULD STATEMENTS

Of all the ways that you can talk negatively to yourself, should statements are the most damaging. Other words that have the same effect are *ought*, *must*, and *have to*. These words judge you. They make you feel guilty and not okay. They are the words of your inner Judge. They come from other people's expectations.

There are three types of shoulds: Having, Doing, and Being. Some examples of each kind follow.

Having Shoulds

You tell yourself that you should have certain things. If you have those things, you will be okay. Maybe you can show others that you are successful. For example:

You should have a good job.
You should have a nice car.
You should have a color TV.

Doing Shoulds

You tell yourself that you should do certain things to be accepted by others. Or, maybe you tell yourself that if you were a good person you would do these things. For example:

You should act your age.
You should go to school.
You should spend time with your children.

Being Shoulds

You tell yourself what traits you need to be okay. You expect yourself to have these qualities in all situations. For example:

You should be nice.
You should be perfect.
You should be strong.

Now think about the shoulds in your life. Make a list of them in the space provided. Then go through the list and for each should decide if it is a Having, Doing, or Being should.

My Shoulds	Kind of Should
________________	________________
________________	________________
________________	________________
________________	________________

If you have lots of shoulds in your life, they control you. You spend your time doing what you should do instead of what you want to do. That's why you need to let go of your shoulds. If you could stop using the words *should*, *ought*, *must*, and *have to* tomorrow, you would instantly feel better about yourself.

By now, using shoulds when you talk to yourself has become a habit. Habits take work to change. Changing negative self-talk to positive self-talk will take work. But you can make a change! Here's how. When you catch yourself saying *should*, *ought*, *must*, or *have to*, change to one of these words:

choose to	consider
prefer	need
want	wish

None of these words are judgmental. Using them will not lower your self-esteem.

CHECKPOINT 2-1

YOUR GOAL: Get 5 or more answers correct.

Read each of the following should statements. Rewrite each statement so it is no longer a should statement. Use words from the list below to rewrite the statements. An example has been completed for you. Give yourself one point for each correct answer.

choose to	consider	prefer
need	want	wish

- I should study more.
  I want to study more.

1. I shouldn't make a mistake.
   ____
2. I should be on time.
   ____
3. I should read better.
   ____
4. I should have called first.
   ____
5. I shouldn't loan Mark money.
   ____
6. I should be more understanding.
   ____

☞ *Check your work on page 47. Record your score on page 52.*

OVERGENERALIZATION

Overgeneralization is when you take one experience and turn it into a rule about life. For example, you apply for a job, and you aren't chosen. You could overgeneralize by saying, "I'll never find a job."

Being passed over for one job does not mean you will not find a job. The overgeneralization, though, just about guarantees you won't find a job. If you believe what you're telling yourself, you'll decide there is no point in trying. You will make your overgeneralization come true by your own actions.

There are several words that are a clue that you are overgeneralizing. They are:

always	all
never	every
everybody	none
nobody	

To turn this type of negative thinking into positive thinking, you need to state only the facts to yourself. No judgment of the facts! Imagine that you and a friend have a fight. Afterwards you can make it worse if you overgeneralize and tell yourself, "Nobody likes me." That isn't even true! But you can make yourself believe it and feel bad.

A better way to talk to yourself is to tell yourself the facts. You could say, "My friend and I had a fight, and he is mad at me now." This statement describes reality. There is no judgment. You can live with the truth. The overgeneralization makes you feel bad about yourself.

CHECKPOINT 2-2

YOUR GOAL: Get 8 or more answers correct.

Read each of the following statements. Underline the clue word or words that tell you the statements are overgeneralizations. Two statements have more than one overgeneralization clue word. An example has been completed for you. Give yourself one point for each correct answer.

- I'll <u>never</u> be able to trust anyone again.

1. None of my bosses have been any good.
2. I always hit every red light.
3. Nobody helps me.
4. Every time I ride the bus, all the seats are taken.
5. Everybody in the class understands this except me.
6. Sara is never on time.
7. All the good jobs are gone.
8. No one works as hard as I do.

Check your work on page 47. Record your score on page 52.

LABELING - YOU DO IT TO YOURSELF

You're twisting reality out of shape again! When you **label**, you describe yourself, others, or situations in a negative way, as in Illustration 2-1. For example, if you make a mistake, you might label yourself by saying, "What a stupid thing to do."

Illustration 2-1

Labels

If someone else makes the mistake, you might label him or her by saying, "What can you expect from a dummy?" If you feel unhappy in a situation you might say, "This is a mess because of those crooked politicians."

In each of the preceding examples, you are using a negative word or phrase. You present the situation as if it is hopeless, and you are powerless to change it. You also upset yourself with the negative labels. How can you feel high self-esteem if you are calling yourself names?

You change labeling the same way you change overgeneralization. You need to describe reality without judging it. Labeling is one more way that you judge. If you make a mistake, you say to yourself, "I made a mistake. How can I fix it?" There is no need to label yourself.

If someone else makes a mistake, you can tell them about it. For example, if the store clerk gives you the wrong change you can say, "You gave me the wrong change." You don't need to label him or her.

Are you beginning to see how many of your bad feelings about yourself come from the way you talk to yourself?

LIFT YOUR SPIRITS

Make a list of some of the negative things you say to yourself. Then rewrite these statements so that they are positive instead of negative. The next time you notice yourself saying the negative statement, change it to the positive one. If you do this for just one month, you will be surprised at how much better you feel about yourself.

FORTUNE TELLING

It would be nice if you could look into a crystal ball and know the future, as in Illustration 2-2. You can't do that, though. Some people act as if they can predict the future. And what they see is usually negative.

Illustration 2-2

Fortune Telling

Fortune telling is predicting the future. It is usually negative. If you use fortune telling, then you may recognize some of these phrases:

Why try? You'll never succeed.
You won't be invited to the party.
They won't hire you for the job.
You don't know enough to pass the test.

Each of these phrases predicts the future. And they don't predict a positive future. If you say these types of statements to yourself often enough, they tend to come true.

If you keep telling yourself you won't get hired for the job you want, you probably won't. Even though you are disappointed when you don't get hired, at least you can say you were right! "See, I told you I wouldn't get hired."

Fortune telling is a vicious circle that keeps you from having high self-esteem. How do you break this habit? Instead of predicting the worst, start planning for what you want. Set a goal for yourself. If you want to pass a test at school, set a goal to get a passing grade. Then ask yourself what you need to do to make that goal come true. Imagine yourself meeting your goal.

LIFT YOUR SPIRITS

Sit down and write yourself a love letter. Tell yourself all the things you like about yourself. Compliment yourself on the things you do well. On days when you feel down, read your love letter to yourself.

Fortune telling and goal setting are very similar *except* that with goals you focus on what you want instead of what you don't want. Any time you notice yourself starting to feel self-doubt, watch out for fortune telling. Avoiding fortune telling is one way to keep your self-esteem high.

CATASTROPHIZING

Catastrophizing is making a mountain out of a mole hill. You turn a headache into a brain tumor. You catastrophize by imagining, "What if. . ." For example, if you make an error at work, you might say to yourself, "What if they fire me?" You blow things out of proportion when you catastrophize. Here are several more examples:

You are driving alone at night and think, "What if I have a flat tire?"

While you study for a test, you think to yourself, "What if I fail?"

You see a TV show about kids and drugs and think, "What if my child is addicted to drugs?"

When you catastrophize, you live in a state of fear. You worry a lot. This needless fear can stop you from doing things that would help you grow. You aren't building your self-esteem.

CHECKPOINT 2-3

YOUR GOAL: Get 4 or more aswers correct.

Read each of the following statements. They are each an example of labeling, fortune telling, or catastrophizing. Write which type of negative thinking each statement is in the space provided. An example has been completed for you. Give yourself one point for each correct answer.

- Some jerk worked on my car, and it still doesn't run right.

  labeling

1. Why ask my boss for the day off? I know he won't give it to me.

   ____

2. If I was a brown-noser, I'd get a raise, too.

   ____

3. You smell gas and think, "What if the house blows up? Where will we live then?"

   ____

4. The landlord won't fix the front step, so why bother telling him about it?

   ____

5. I'm a worthless excuse for a parent.

   ____

6. What if this assignment isn't perfect, and I flunk out of school?

   ____

☞ *Check your work on page 48. Record your score on page 52.*

The secret to avoiding negative self-talk is to describe reality. Don't judge it. Judgment kills your self-esteem. Try to tell the truth about what you see and observe. Avoid giving your opinion if it will be a judgment. Later in this book you will learn more ways to replace your negative self-talk with positive self-talk.

It will not be easy to change your habits of negative self-talk. But you can do it! Don't criticize yourself when you notice yourself in an old habit. Instead, congratulate yourself for noticing the negative self-talk. You can't change something until you are aware of it. As you practice the ideas in this unit, you will notice that less and less of your self-talk is negative.

LIFT YOUR SPIRITS

One way to break the negative thinking habit is to do something called Thought Stopping. When you notice a negative thought, tell yourself STOP! and picture a red stop sign. You will stop the negative thought. Then change your negative thought into one that is positive. Use the suggestions in this unit to turn a negative thought into a positive one.

WHAT YOU HAVE LEARNED

You talk to yourself all day long. Some of what you say is positive, and some of what you say is negative. The negative self-talk includes shoulds, overgeneralizations, labeling, fortune telling, and catastrophizing. All of these negative thoughts are judgments of you. They lower your self-esteem. You want to observe and describe reality instead of judging it with these negative thoughts.

PUTTING IT TOGETHER

ACTIVITY 2-1 YOUR GOAL: Get 8 or more answers correct.

Read each of the following statements. Write *T* if a statement is true in the space provided. Write *F* if the statement is false. An example has been completed for you. Give yourself one point for each correct answer.

- __T__ You talk to yourself all day long.

1. ______ Overgeneralization is the most damaging type of self-talk.
2. ______ There are three types of shoulds — wishing, wanting, and having.
3. ______ Negative self-talk is judgmental.
4. ______ If you say words like never, always, no one, or everybody, you are overgeneralizing.
5. ______ You can change negative self-talk to positive self-talk.
6. ______ You can change your shoulds by using the words choose, prefer, want, need, consider, or wish.
7. ______ Negative self-talk is a habit that can be changed.
8. ______ Fortune telling and goal setting are very similar.
9. ______ Fortune telling is when you use the words *should*, *ought*, *must*, or *have to*.
10. ______ Your self-talk affects how you feel about yourself.

☞ ***Check your work on page 48. Record your score on page 52.***

ACTIVITY 2-2 YOUR GOAL: Get 10 or more answers correct.

Read the following paragraph. Underline any negative thought. Write the type of negative thought above the word or phrase. Give yourself one point for each word or phrase that is correctly underlined. Give yourself another point if you correctly stated the type of negative thought.

I got up late this morning. I should have gone to bed earlier. The bus is never on time, so it really didn't matter. All the way to school I worried about my math test. "What if I fail it? What if I flunk the course?" I'm a real dummy when it comes to math. I've never understood fractions, and I probably never will.

☞ ***Check your work on page 48. Record your score on page 52.***

UNIT 3

Positive Self-Talk

WHAT YOU WILL LEARN

When you finish this unit, you will be able to:

- Create your own Fan Club.
- Write an affirmation.

As you've learned, the Judge uses negative self-talk to keep your self-esteem low. In this Unit, you are going to meet your Fan Club and learn how to use positive self-talk.

MEET YOUR FAN CLUB

What is a Fan Club? It's a group of people who believe in you 100 percent. In their eyes, you can do no wrong. They're proud of your every achievement, no matter how big or small. According to your Fan Club, if you face a disappointment, it's because there's something better waiting for you. When you feel uncertain, your Fan Club encourages you. When life feels hard, your Fan Club reminds you how much you're learning. They are a very positive group, as you can see in Illustration 3-1. They are with you all the time to help you love yourself and be the best you can be.

Illustration 3-1

Your Fan Club

Each person has a Fan Club inside of him or herself. Sometimes you're so busy listening to the Judge that you don't pay attention to your Fan Club. Who you listen to—the Judge or your Fan Club—controls your level of self-esteem.

Who are the members of your Fan Club? You decide. There is only one test to pass to be in your Fan Club. *Does he or she believe in you?* A Fan Club is no place for someone who doesn't believe that you are wonderful!

Some people have their mother, father, or both in their Fan Club. Other people have a favorite instructor. Many people have Jesus or God in their Fan Club. Friends, children, bosses, and co-workers have been found in Fan Clubs. Even pets have been included. You can have as many people (or animals!) as you want in your Fan Club.

Stop now and think about who is in your Fan Club. Don't forget yourself, either! You can root for you! Write down the names of the people in your Fan Club in the space provided.

My Fan Club:

____

____

____

Anytime you feel down, you can ask your Fan Club to cheer you. Anytime the Judge starts talking to you, you can decide to listen to your Fan Club instead. You aren't alone. Your Fan Club is on your side.

START USING AFFIRMATIONS

Your Fan Club uses a special kind of self-talk called affirmations. An **affirmation** is a positive statement you say to yourself that builds you up, rather than tearing you down. An affirmation is a way of talking to yourself that says you count. You're important. You're okay.

An affirmation has four key characteristics: it is positive, said in the present tense, visualized, and repeated often. If any one of the characteristics is missing, the affirmation is not as powerful. Let's look at each of the parts in turn.

An Affirmation Is Positive

You might be thinking to yourself, "Well of course!" But it is not as easy as it sounds. For an affirmation to be positive, it needs to say what you *want*, not what you *don't* want.

Look at the following example Rick wrote:

"I don't want to get sick."

LIFT YOUR SPIRITS

There is a well-known children's story about the Little Engine That Could. In the story, a little engine needs to pull a heavy load over a mountain. He used affirmations. He first told himself, "I think I can, I think I can." Then he said "I know I can, I know I can." The next time you face a hard task, remember the words of The Little Engine That Could.

He thought this was a positive statement. It's not a negative statement, but it doesn't meet the rule to be a positive statement, either. "I don't want to get sick," tells us what Rick *doesn't* want, but we don't know what he *does* want. Here's how he could rewrite it to be positive:

"I am healthy."

You may know what you don't like or don't want. It is much harder to know what you *do* like and want. Your Fan Club needs to know specifically how to cheer you on. They get excited about saying, "Rick is healthy!" They don't think "Rick doesn't want to get sick" is motivating.

The key to making your affirmation positive is avoiding the word *not*. Watch out for do not (don't) and will not (won't). Replace these words with what you want.

Here's another example: "I don't make mistakes." Do you see the problem for your Fan Club? They don't know what this means. It isn't positive. To make it a powerful affirmation, take out the "don't." Decide what you want. One way to change it would be to say, "I get A's on my math tests." Another affirmation might be, "I check my work." Each of the last two examples are positive. They say what you want.

CHECKPOINT 3-1

YOUR GOAL: Get 5 answers correct.

Read each of the following statements. Decide if they meet the first step to be an affirmation. Write *yes* in the space provided if a statement is positive. Write *no* if it is not. An example has been completed for you. Give yourself one point for each correct answer.

- __no__ I don't waste time.

1. ________ I manage my time.
2. ________ I don't owe anyone any money.
3. ________ I won't be late.
4. ________ I'm not lonely.

5. ________ I love my job.

6. ________ I'm not scared.

☞ *Check your work on page 48. Record your score on page 52.*

An Affirmation Uses the Present Tense

The second characteristic of an affirmation is that it is said in the present tense. Not the past tense. Not the future tense. Say it as if it is true NOW. Today.

Stating an affirmation as if it's true is hard for some people to do. If it isn't true yet, it seems like you are lying to yourself. What you're really doing is using a special feature of the mind.

By stating your affirmation as if it is true right now, you give yourself courage to do something today to make it true. You start to think about yourself in a different way. After all, when you use negative self-talk, don't you put it in the present tense? Don't you say things to yourself like, "I'm so stupid," rather than, "I'm going to be stupid." You know how powerful the negative self-talk is. You're applying the same rules to your positive self-talk. So instead of saying, "I will have confidence," say, "I have confidence." Here are more examples:

FUTURE TENSE	PRESENT TENSE
I will do a good job.	I do a good job.
I will pay the bills on time.	I pay the bills on time.

The point is, you want to affirm to yourself what you want, as if it is true today. Put your affirmations in the present tense. Decide to believe them. This sets in motion all kinds of mental processes that will make them come true.

CHECKPOINT 3-2

YOUR GOAL: Get 5 answers correct.

Rewrite the following affirmations so that they are in the present tense. An example has been completed for you. Give yourself one point for each correct answer.

- I will get healthy.
  I am healthy.

1. I will lose weight.

   ____

2. I will pass the test.

   ____

3. I will get a new job.

____

4. I will study for the next test.

____

5. I will read to the kids before they go to bed.

____

☞ *Check your work on page 48. Record your score on page 52.*

Visualize Your Affirmation

When you **visualize** something, you see a picture of it in your mind. You make pictures in your mind all day long. Everybody does. When you talk to someone, you "see" what they say. When you think about something, you often "see" what you're thinking about. Few people realize that you can also visualize high self-esteem!

To explain how you can use visualization to raise your self-esteem, you need to understand how the mind works. The mind is an amazing tool. Yet, it cannot tell the difference between imagination and reality.

Try this experiment. Imagine walking into your kitchen. You pick up a large yellow lemon. Feel the slightly rough texture of the skin against your hands. Now, get a paring knife and cut the lemon in half. Notice how juicy the lemon is as you cut it. In fact, it squirts you! Now smell the fresh, clean smell of the lemon. Bring half of the lemon up towards your nose. Breathe in. Doesn't it smell good? Now, take a big bite of the lemon. Sink your teeth into the fruit....

What is happening in your mouth right now? Is it watering? If so, you might wonder why. You aren't eating a lemon. Or are you? In your mind you are. Your mind reacts to anything it can picture clearly, and in detail, as if it is real. That's why you can salivate just thinking about eating a lemon.

If it can work with a lemon, why not make it work for your self-esteem? Maybe you've been picturing yourself as not being very lovable. Maybe you've pictured yourself as a poor student. Maybe

LIFT YOUR SPIRITS

An American pilot was a prisoner of war (POW) in North Vietnam for many years. He lost 80 pounds and much of his health while he was a POW. When he came home, one of the first things he wanted to do was play a game of golf. He played a great game. No one could understand it because he was so thin and he hadn't played in years. He told them that while in prison he visualized himself playing golf every day.

you've pictured yourself in a job you don't like. These are the negative pictures that come from the Judge. The Judge doesn't paint pretty pictures in your mind. But because the pictures seem so real, *you act as if they are true*. Then you feel bad about yourself and have low self-esteem.

You can change the pictures! Instead of listening to the Judge, listen to your Fan Club. Start creating some new pictures in your mind. Your life won't change overnight, but it will start to change. You can start visualizing yourself today as the kind of person you want to be. Replace the old pictures with new, positive ones. Listen to your Fan Club. "See" what they have to say to you. You can choose which inner voice you will listen to. You can decide who will be in charge of your life — the Judge, or your Fan Club. It's up to you.

This is an important point. *What you see is what you get*. Until you can see yourself in the kind of job you want in your mind's eye, you won't get the kind of job you want. Until you can see yourself as a confident person in your mind, you won't be a confident person. Until you can see yourself having friends in your mind, you won't have friends. Be careful about the movies that you watch in your mind! They affect how you feel about yourself. They affect who you are.

CHECKPOINT 3-3

YOUR GOAL: Get 10 points.

Think about the way the Judge talks to you. Write a paragraph about the person you picture when you listen to the Judge. Then think about your Fan Club. Write a paragraph about the person you picture when you listen to your Fan Club. An example has been completed for you. Give yourself five points for each paragraph.

- Who do you see when you listen to the Judge?

  I see a person who can't do anything right, so she doesn't try new things. People laugh at her. She doesn't like herself. She feels inadequate most of the time.

  Who do you see when you listen to your Fan Club?

  I see a confident person who can do anything she decides to do. She takes risks, has friends, and feels happy. She learns new things.

a. Who do you see when you listen to the Judge?

____

____

____

____

b. Who do you see when you listen to your Fan Club?

____

____

____

____

☞ *Check your work on page 48. Record your score on page 52.*

Say Your Affirmations Often

Self-talk goes on all the time. In the past, you have listened to the Judge most of the time. Now it is time to listen to your Fan Club.

Here's how you do it. First, when the Judge starts talking, decide you aren't going to listen. Talk back to the Judge with words from your Fan Club. It is a good idea to have some phrases in mind that you can use. For example, you could say one of these statements to the Judge:

I like myself.
I am special.
I am somebody.
I count.
I am important.

Second, pick two or three affirmations that you will say to yourself every morning when you get up and every night before you go to bed. You can also say the affirmations to yourself during the day. Think of these as being self-esteem food. Each day you eat food to keep your body physically fit. You know that if you eat too many sweets, drink too much, or skip meals, your body doesn't feel good. You need to give it healthy food so it can do it's best.

You also need to feed your self-esteem. If the only food it gets is from the Judge, you won't feel good about yourself. Decide to feed your self-esteem at least twice a day. Here are some examples of affirmations that you can use.

I love and accept myself.
I'm getting better and better everyday.
I now release all comparisons with others.
I am at peace.

I love and appreciate all of my natural abilities.
I let go of negative thoughts.
I am a loving person.

You don't need to feel limited to these affirmations. You can write your own. You now know all the parts to include.

CHECKPOINT 3-4

YOUR GOAL: Get 4 or more points.

Write two affirmations you can use when the Judge starts to talk to you. Then write two affirmations that you can use each day to build your self esteem. Review the last section for examples. Give yourself two points for each one you write.

Affirmation 1: ______________________________

____

Affirmation 2: ______________________________

____

Affirmation 3: ______________________________

____

Affirmation 4: ______________________________

____

☞ *Check your work on page 48. Record your score on page 52.*

Keep your affirmations short. You want to be able to remember them! Don't try to say too many of them at one time. It works best to choose two or three that you will say to yourself each day. Begin by using affirmations to raise your self-esteem. As you succeed in that area, you can start using affirmations in other parts of your life. There is no limit to their possible use.

Affirmations are a tool you can use to change your life. Test their power by using them every day for thirty days. You'll like what you discover.

WHAT YOU HAVE LEARNED

You have a Fan Club! You can choose to listen to the Judge and lower your self-esteem. Or you can listen to your Fan Club and build your self-esteem. Your Fan Club will talk to you in affirmations. To be most powerful, an affirmation is positive, said in the present tense, visualized, and used often.

ACTIVITY 3-1

YOUR GOAL: Get 6 or more points.

Answer the following questions in the space provided. An example has been completed for you. Give yourself two points for each answer.

- What is a Fan Club?

  A group of people in your mind who believe in you. They cheer you on and talk positively to you.

1. Name and explain the first part of an affirmation.

   ____

   ____

   ____

2. Name and explain the second part of an affirmation.

   ____

   ____

   ____

3. Name and explain the third part of an affirmation.

   ____

   ____

   ____

4. Name and explain the fourth part of an affirmation.

   ____

   ____

   ____

☞ *Check your work on page 48. Record your score on page 52.*

ACTIVITY 3-2 YOUR GOAL: Get 8 or more answers correct.

The following statements are not affirmations as written. Why not? Write the reason in the space provided. An example has been completed for you. Give yourself one point for each correct answer.

- I don't do well on tests.

  Negative ______

1. I won't be chosen for the job.

   ______

2. I will learn to accept myself.

   ______

3. I won't get any help.

   ______

4. I wish I could read better.

   ______

5. I will stand up for myself.

   ______

6. I don't do well in job interviews.

   ______

7. I won't be liked.

   ______

8. I won't lose my keys.

   ______

9. I won't be turned down for the raise.

   ______

10. I will be happy.

    ______

☞ *Check your work on page 48. Record your score on page 52.*

UNIT 4 I Like Myself!

WHAT YOU WILL LEARN

When you finish this unit, you will be able to:

- Keep your self-esteem high.
- Build the self-esteem of others by affirming others.
- Set SMART goals.

Liking yourself is a lifetime job. Once you have mastered the ideas in this book to build your self-esteem, you aren't done. Your high self-esteem needs to be maintained.

MAINTAIN HIGH SELF-ESTEEM

From time to time your self-esteem will slip. Your self-esteem is like a yo-yo. It goes up and down. You want to keep it high as much as you can. Here are some tips to maintain high self-esteem.

Spend Time With Positive People

As you've learned, other people can help you build your self-esteem. Positive people can also help you maintain your self-esteem.

You decide who your friends are. Choose people who believe in you, like in Illustration 4-1. Choose people who look for the positive in even a negative situation. On days when you feel your self-esteem is low, call one of these people to lift your spirits.

Stop spending time with people who bring you down. At work, take breaks and eat lunch with positive people. Your self-esteem is too important to let anyone damage it. You may need to change some of your relationships to keep your self-esteem high.

Illustration 4-1

A Friend Believes in You

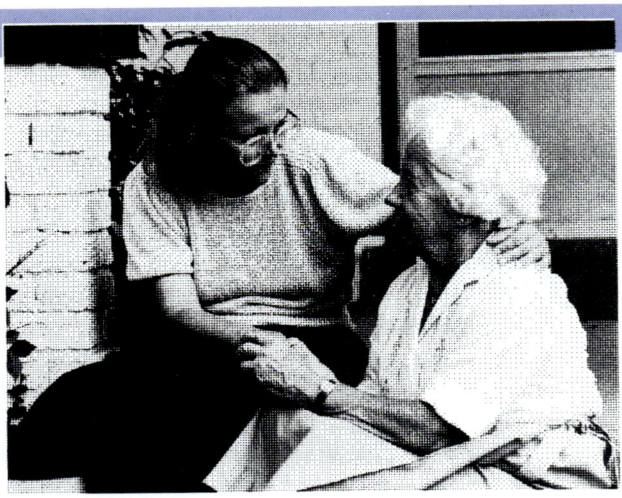

Look To The Future, Not The Past

Anyone can look back and see a better way to do things. It is easy to get stuck thinking about what "could have been" or "should have been." The problem with this type of thinking is that you can't change the past.

However, you can *learn* from your past. You can use the past to change how you act in the future. You can keep your self-esteem high if you use today to make a better tomorrow. Accept that you are a developing human being. Each mistake is a chance to learn. When things don't turn out the way you had hoped, decide how you will do it differently the next time.

Keep yourself focused on the future, when you can change your behavior. Notice the difference in the following examples between looking back and looking forward.

PAST: "I judged myself again. I should have accepted my mistake. I just can't get it right."

FUTURE: "I realize I made myself feel bad when I judged myself. Next time I will describe the situation instead of judge it."

PAST: "I don't know why I didn't speak up when my supervisor asked me to work late. I could have saved myself a lot of problems if I would have said how I felt."

FUTURE: "I notice I didn't speak up. It is still hard for me to express my feelings. I'm going to practice saying what I feel so I'm not in this situation again."

PAST: "I feel so awful about yelling at the kids. I shouldn't have said the things I said to them."

FUTURE: "I lost control of my feelings with the kids. Next time I feel mad I'm going to send them to their room until I cool off."

In each example, the first statement judged the past and caused feelings of low self-esteem. There is no way to change the situation or fix the problem. You are stuck when you look back.

The second statement looks to the future with a plan to do better. Describe your past with a focus on what you will change in the future.

CHECKPOINT 4-1

YOUR GOAL: Get 4 or more answers correct.

Read each of the following statements. Decide if each looks to the past or the future. Write *past* or *future* in the space provided. An example has been completed for you. Give yourself one point for each correct answer.

- ● past I shouldn't have said that.
1. ______ Why did I spend money on those magazines?
2. ______ Next month I'll make a budget so I don't run short on money.
3. ______ Next time I buy a new appliance I'll ask more questions.
4. ______ I knew I should have brought an umbrella.
5. ______ If I'd left earlier I wouldn't be running late.

☞ *Check your work on page 48. Record your score on page 52.*

Accept Yourself

No one can be a better YOU than you! If you try to be someone else, you will fail. And you will lower your self-esteem in the process. The world needs *you* to be *you*. No one else can do it. On the days when you feel your self-esteem starting to slip, say to yourself, "I am me. There is no one else quite like me. I will be the best me that I can be."

You may set goals to develop yourself. That doesn't change the fact that you are okay today as you are. You don't criticize a seed because it isn't a tree yet. Accept yourself. Like a seed, you have lots of growing to do. That's normal. Everyone needs to grow and develop him or herself. Your life is a process of becoming the best you can be. Accepting yourself as you are now, today, makes it possible for you to grow and develop. When you feel okay about yourself, you are able to risk change.

Do Something You Love

You are like a battery. You have energy that you give to others all day long. Like a battery, you sometimes need to be recharged. Being with positive friends is one way to charge your battery. Another way is to do something you love to do.

What do you love to do? Make a list of ten things right now. Put things on your list that you can do alone, and things you can do with a friend. List things you can do for free, as well as things that cost money.

Things I love to do:

____

____

____

____

How often do you do the things you love? Look at this list when your self-esteem slips. Then choose something from the list to do to give your self-esteem a lift.

CHECKPOINT 4-2

YOUR GOAL: Get 4 or more answers correct.

Read each of the following statements. If a statement is true, put a *T* in the space provided. If it is false, put an *F* in the space provided. An example has been completed for you. Give yourself one point for each correct answer.

- ___F___ If your self-esteem slips, there's nothing you can do about it.

1. ______ If you have a hard day, it can help to spend time with positive people.
2. ______ You will feel better about yourself if you try to be like someone else who has high self-esteem.
3. ______ Most people who look back into the past see things they wish they had or hadn't done. This lowers self-esteem.
4. ______ You can lower your self-esteem by doing things you love.
5. ______ The keys to maintaining high self-esteem are to be around positive people, look to the future, be yourself, and do things you love.

☞ ***Check your work on page 49. Record your score on page 52.***

AFFIRM OTHERS

You are not alone in wanting high self-esteem. Everyone wants to like him or herself. As you start to affirm yourself, you'll find that it feels natural to affirm others, too.

Notice What People Do That You Like

So often we look for what people do wrong. That is the Judge inside who wants to criticize. As you listen more to your Fan Club and less to your Judge, you'll notice that you don't judge yourself *or others* as often.

Look for what people do right. Don't look for big things. Look for the little things. The coworker who greets you with a smile. The friend who listens to you. The family member who helps out. The neighbor who lets you get on the bus first. The stranger who holds a door open for you. These small kindnesses make a difference in your life.

As you put your attention on these gestures of goodwill, you will notice more and more of them. The world will seem a nicer place. Nothing has changed, except your attitude.

Say Thank You

Take a minute to say "Thank you" when someone does something you like. The more specific you can be, the more it will mean to the other person. When you are specific, the person knows exactly what you liked. They don't have to guess. For example, "Thanks for your help" doesn't say as much as "Thank you for helping me fold the sheets."

Another way to say thanks is to use the words "I appreciate." Again, follow the words with as much detail as possible. "I appreciate your thoughtfulness" doesn't mean as much as "I appreciate you calling to ask how my Dad is feeling." When you are specific, the other person knows that you noticed. A "Thank you" or "I appreciate. . . ." builds self-esteem in others. When they know you noticed, they are more likely to do the same thing again. That's good for both of you!

Don't Judge Others

It is so easy to judge others. This hurts your self-esteem and their self-esteem. As long as you judge others, it is a sign that you are judging yourself. Here are some things to avoid.

Don't put others down. How will a put-down help you or the other person? It won't. When you feel good about yourself at someone else's expense, it is a sign of low self-esteem. Instead of putting others down, challenge yourself to find something to appreciate. You'll both feel better.

Don't gossip. If you hear gossip, let it stop with you. Don't pass it on. If someone wants to talk behind another person's back, say, "I don't want to hear it."

LIFT YOUR SPIRITS

"If the only prayer you ever say is Thank You, it's enough."

Meister Eckhart

Don't laugh at sexist or racist jokes. Those types of jokes come from low self-esteem and comparisons. If you won't be an audience, the jokes will stop.

Don't take advantage of people with low self-esteem. Be a person who helps them build their self-esteem.

CHECKPOINT 4-3

YOUR GOAL: Get 4 or more answers correct.

Read each of the following situations. Decide how you would respond to build self-esteem in another. Write your answer in the space provided. An example has been completed for you. Give yourself one point for each correct answer.

- When it was raining your spouse said, "You wait here. I'll get the car and pick you up."

  I appreciate you bringing the car to pick me up so I didn't get wet.

1. A co-worker wants to tell you about the trouble Kelly is in.

   ____

   ____

2. You're with a group of friends after work, and they start to tell sexist and racist jokes.

   ____

   ____

3. April brings you a can of fruit juice at break time.

   ____

   ____

4. You call a store about the price of an item. The person who answers the phone is friendly and gets you an answer quickly.

   ____

   ____

5. You have a doctor appointment at 2:00 p.m. The doctor's office calls you at 1:15 to say the doctor is running late and asks if you can come at 2:30 p.m.

   ____

   ____

Check your work on page 49. Record your score on page 52.

SET SMART GOALS

You have learned many things that will help you build your self-esteem. Once you decide which ideas you want to use, you need to set goals and take action. Knowledge without action is like a car without gas. You won't get too far.

Most people don't set goals. Of the ones who do, few know how to set a goal so that they will reach it. If the only idea you get from this book is to set goals — and you do — your life will make a change for the better.

All you need to remember is to set SMART goals. Here's what SMART stands for:

S - SPECIFIC
M - MEASURABLE
A - ACHIEVABLE
R - REALISTIC
T - TIMEBOUND

Be Specific

You want your goal to be as clear as possible. Then you know what you want to achieve. If a goal is vague, or too general, it won't help you get what you want.

For example, if you had the goal, "Lose Weight," it wouldn't work well. Why not? It isn't specific. To make the goal specific, you could say, "I want to lose ten pounds." Here are some other examples.

VAGUE	SPECIFIC
Be a better parent.	Read to my child fifteen minutes each night.
Get better grades.	Earn a B in math class.
Have a nice car.	Drive a car with no rust on the body.
Look better.	Get a new hairstyle.

If the goal isn't specific, you don't have a way to tell if you are meeting it or not. If you can't measure your progress, you will probably drop the goal.

LIFT YOUR SPIRITS

A visitor was watching two stonemasons work. He went up to the first stonemason and asked what he was doing. The reply was, "I'm cutting stone." Then the visitor went up to the second stonemason and asked the same question. The reply was, "I'm building a cathedral." Which stonemason had a goal? Do you have goals?

Make Goals Measurable

You need to know if you are meeting your goals. You need a way to measure them. If you can't tell if you are reaching them, why set them?

You can't measure a goal that is vague. Look at the preceding examples of goals. How would you know if you were a better parent? There is no way to tell. But would you know if you read to your child fifteen minutes each night? Yes. That is a measurable goal. Would you know when you lost 10 pounds? Yes. Could you tell if you got a new hairstyle? Yes. A specific goal will give you a way to measure your progress. You will know when you reach the goal. When you do, you can celebrate.

Be sure you have the means to measure your goal. For example, you can measure weight loss with a scale. But do you have a scale? If not, the goal isn't measurable for you. You might need to change it to say, "Fit into a size 10 dress," if you are a woman. Or "Wear pants with a 38-inch waist," if you are a man. You could measure your dress or waist size when you go to the store to buy clothes. It isn't enough that the goal can be measured. *You* need to be able to measure the goal.

CHECKPOINT 4-4

YOUR GOAL: Get 4 or more answers correct.

Read each of the following goals. Decide if each goal is specific and measurable. If it is, write *yes* in the space provided. If it isn't write *no*. An example has been completed for you. Give yourself one point for each correct answer.

- yes Pay all of my bills on the day they are due or earlier.

1. ______ Get a better job.
2. ______ Get married.
3. ______ Spend more time with friends.
4. ______ Watch less TV.
5. ______ Get up when the alarm goes off.

☞ ***Check your work on page 49. Record your score on page 52.***

Is Your Goal Achievable?

The preceding question is very important if you tend to be a perfectionist. Has what you want to do ever been done before? Is it possible? Do you know how to do it? Some perfectionists set goals for themselves that aren't achievable. Then, when they don't achieve the goal, they feel bad about themselves.

You may decide that your goal is achievable, but you need to reach some other goals first. For example, it may be achievable for you to read to your child, but you may need to develop your reading skills first.

Is Your Goal Realistic?

Being realistic about your goal is important, too. You need to know yourself well enough to know what you can and can't do. There are many goals that are achievable, yet *unrealistic* for you.

Let's go back to the example of reading to your child. It may be achievable, but what if you work second shift? You may not be home when your child goes to bed. The goal isn't realistic. It needs to be changed.

If you are on a medicine that causes you to gain weight, it may not be realistic to lose weight while you are on the medicine. If you want to buy a "new" used car without rust, and you live in Buffalo where there is snow and salt trucks, your goal may not be realistic.

You will need to decide what is realistic for you. Knowing your talents and skills will help you set goals that are realistic for you.

It is okay to make a smaller goal than you planned at first. You can make another goal after you achieve the first one. It is also okay to break a big goal into smaller goals. As you reach each of the smaller goals, you will be closer to your big goal.

Make Goals Timebound

A goal without a deadline is only a dream. Dreams don't often come true. Decide when you want to reach your goal. Set a date. It is a well-known fact that most people are motivated by a deadline. Even your bills have a due date to motivate you to pay them.

If you want to lose ten pounds, then set a date. Maybe you think you can lose one pound per week. Then count out ten weeks from today. That is your deadline.

Perhaps you want a new hairstyle for summer. Then set a date to have a new hairstyle by June 1. If you want a new car, when are you going to get it? Next month? Next year? In five years? You decide. Put deadlines on your goals. Without a timeframe, you probably won't reach your goals.

One more point about goals. Write them down. Put them in a place where you can see them often. A bathroom mirror works well for many people. Other people write them on a slip of paper and carry them in a purse or pocket. If you can see your goals everyday, it will remind you of what is important. Before you watch TV, glance at your goals. You might decide to study for your math test instead!

CHECKPOINT 4-5

YOUR GOAL: Get 4 or more answers correct.

Read the following goals. Make any needed changes so that they are SMART goals. An example has been completed for you. Give yourself one point for each correct answer.

- Build my self-esteem.

  Write two affirmations by the end of the week.

1. Fix my teeth.

   ____

2. Make new friends.

   ____

3. Mow the lawn.

   ____

4. Go back to school.

   ____

5. Move to a bigger apartment.

   ____

☞ ***Check your work on page 49 . Record your score on page 52.***

Only you can decide how you will use what you've learned in this book. No one can choose your goals for you. Choose your goals based on what is important to you. Then you will want to reach your goals because *you* set them.

There are many goals you can set to build your self-esteem and keep it high. Don't let this be the end of the book. Make it the beginning of taking action. Good luck!

WHAT YOU HAVE LEARNED

From time to time your self-esteem will slip. You can keep it high by being with positive people, accepting yourself, looking to the future, and doing things you love. You can make a difference to others if you take time to affirm them. Look for what other people do well and let them know that you noticed. To grow and develop you will need to set goals for the changes you want to make. Set SMART — specific, measurable, achievable, realistic, and timebound — goals.

PUTTING IT TOGETHER

ACTIVITY 4-1 YOUR GOAL: Get 4 points.

Write two goals for yourself to build your self-esteem in the space provided. B sure they are SMART goals. An example has been completed for you. Giv yourself two points for each goal.

- I will say two affirmations a day for the next thirty days.

Goal 1: ______________________________

____

Goal 2: ______________________________

____

☞ *Check your work on page 49. Record your score on page 52.*

ACTIVITY 4-2 YOUR GOAL: Get 8 or more answers correct.

Answer each of the following questions in the space provided. An example has been completed for you. Give yourself one point for each correct answer. Check your work.

- Why is it important to spend time with positive people?

  Positive people help you build and maintain high self-esteem.

1. List three ways to maintain high self-esteem.

   ____

   ____

2. What does it mean to "look to the future, not the past?"

   ____

   ____

3. Change the following statement from looking to the past to looking to the future. "I should have studied for the test. I would have done better."

   ____

   ____

4. How can you affirm another person?

   ____

   ____

5. Name two times when it is important to be specific.

____

____

6. Give one example of the words you can use to begin an affirmation to another person.

____

____

7. Give two examples of how to avoid judging others.

____

____

8. What do the letters SMART stand for?

____

____

9. Why do goals need deadlines?

____

____

10. What is one reason to write out your goals?

____

____

☞ ***Check your work on page 49. Record your score on page 52.***

CHECKING WHAT YOU LEARNED

Now you can see how much you have learned about building your self-esteem. These 25 questions cover the main topics you studied in this book. There is no time limit, so take your time.

After you finish, check your answers. Give yourself 2 points for each correct answer. Record your score on your Personal Progress Record. The evaluation chart will tell you where you may need additional study.

DIRECTIONS: Read each of the following statements. If a statement is true write a *T* in the space provided. If it is false write an *F* in the space provided.

1. ____ Self-esteem is what other people think of you.
2. ____ Being around someone with low self-esteem makes you feel good about yourself.
3. ____ People with low self-esteem are often negative.
4. ____ Your self-image is addicted to the opinions of other people.
5. ____ People who feel worthless have high self-esteem.
6. ____ Self-image is what you want other people to think about you.
7. ____ Self-talk is always negative.
8. ____ When you use the word *should* on yourself, you are judging yourself.
9. ____ "You never study" is an example of a should statement.
10. ____ The fewer shoulds you use, the better you will feel about yourself.
11. ____ To overgeneralize means to take one experience and turn it into a rule about life.
12. ____ Fortune telling is when you predict the future.
13. ____ Catastrophizing is making a mountain out of a mole hill.
14. ____ Only people with high self-esteem have an inner Fan Club.
15. ____ An affirmation is a positive statement you say to yourself.
16. ____ Visualization is when you picture something in your mind.
17. ____ "I'm not lonely," is an example of an affirmation.
18. ____ Affirmations are stated in the past tense.
19. ____ "I will do good work," is an example of an affirmation.
20. ____ The more you use affirmations the better you will feel about yourself.
21. ____ Positive people help you maintain high self-esteem.
22. ____ If you use the past to change how you behave in the future, it becomes a tool for building self-esteem.
23. ____ To build the self-esteem of others, tell them when they do something you like.
24. ____ A SMART goal is measurable.
25. ____ "Get better grades" is an example of a SMART goal.

☞ ***Check your work on page 49. Record your score on page 53.***

GLOSSARY

A

Acceptance The opposite of judgment.
Affirmation A positive statement you say to yourself that builds you up rather than tears you down.

C

Castastrophizing Making a mountain out of a mole hill.

F

Fortune Telling Predicting the future. It is usually negative.

J

Judge A little voice inside of you that tells you all the things you are doing wrong.
Judgment Assigns value, says something is good or bad, right or wrong.

L

Label To describe yourself, others or situations in a negative way.

O

Overgeneralization When you take one experience and turn it into a rule about life.

S

Self-acceptance When you can say to yourself, "I am okay."
Self-confidence Believing in yourself.
Self-esteem How you feel about yourself.
Self-image What you want other people to think of you.
Self-respect The opinion you have about yourself.
Self-talk Conversations that you have with yourself.
Self-worth Your belief that you are a valuable human being.

V

Values Your beliefs about what is right.
Visualize To see a picture of something in your mind.

INDEX

ANSWERS

✔ CHECKING WHAT YOU KNOW

1. T
2. T
3. T
4. T
5. T
6. T
7. F
8. F
9. T
10. F
11. T
12. F
13. F
14. T
15. F
16. T
17. F
18. F
19. T
20. T
21. F
22. F
23. T
24. F
25. F

UNIT 1

CHECKPOINT 1-1, page 3

1. High
2. Low
3. Middle
4. Low
5. High
6. Middle
7. Low
8. High

CHECKPOINT 1-2, page 5

1. Values
2. Self-respect
3. Self-worth
4. Self-worth, self-respect
5. Self-esteem

CHECKPOINT 1-3, page 7

1. Esteem
2. Image
3. Esteem
4. Esteem
5. Image
6. Esteem
7. Image
8. Image
9. Esteem
10. Image

ACTIVITY 1-1, page 9

1. Low
2. Middle
3. Image, Self-image
4. Self-esteem
5. Self-image
6. Self-esteem
7. Self-worth
8. Image
9. Depends
10. Values

ACTIVITY 1-2, page 9

1. a
2. b
3. a
4. b
5. a
6. c
7. b
8. c

UNIT 2

CHECKPOINT 2-1, page 13

These are sample statements. You may have used a different word to rewrite the statement.

1. I prefer not to make mistakes.
2. I choose to be on time.
3. I want to read better.
4. I wish I had called first.
5. I choose not to loan him money.
6. I want to be more understanding.

CHECKPOINT 2-2, page 14

1. <u>None</u> of my bosses have been any good.
2. I <u>always</u> hit <u>every</u> red light.
3. <u>Nobody</u> helps me.
4. <u>Every</u> time I ride the bus, <u>all</u> the seats are taken.
5. <u>Everybody</u> in the class understands this except me.
6. Sara is <u>never</u> on time.
7. <u>All</u> the good jobs are gone.
8. <u>No one</u> works as hard as I do.

CHECKPOINT 2-3, page 17

1. fortune telling
2. labeling
3. catastrophizing
4. fortune telling
5. labeling
6. catastrophizing

ACTIVITY 2-1, page 20

1.	F	6.	T
2.	F	7.	T
3.	T	8.	T
4.	T	9.	F
5.	T	10.	T

ACTIVITY 2-2, page 20

I got up late this morning. I should have gone to bed earlier. (should statement) The bus is never on time, (overgeneralization) so it really didn't matter. All the way to school, I worried about my math test. "What if I fail it? What if I flunk the course?" (catastrophizing) I'm a real dummy (labeling) when it comes to math. I've never (overgeneralization) understood fractions and I probably never will. (fortune telling)

UNIT 3

CHECKPOINT 3-1, page 23

1. Yes
2. No
3. No
4. No
5. Yes
6. No

CHECKPOINT 3-2, page 24

1. I weigh my ideal weight.
2. I pass tests.
3. I have a new job.
4. I study for tests.
5. I read to the kids before they go to bed.

CHECKPOINT 3-3, page 26

Ask your instructor, a fellow student, friend or family member to read what you wrote. Give yourself 5 points for each paragraph that they agree describes how a Judge or Fan Club might see you.

CHECKPOINT 3-4, page 28

Ask your instructor or a fellow student to read each affirmation you wrote. Give yourself 2 points for each affirmation if he or she agrees that it is positive and written in the present tense.

ACTIVITY 3-1, page 29

1. An affirmation is positive. This means say what you want, not what you don't want. Avoid the use of the word not.
2. Affirmations use the present tense. This means don't use the future or past tense.
3. Visualize the affirmation. This means picture your affirmation in your minds eye.
4. Say your affirmation often. This means talk back to the Judge and say your affirmations every morning and night.

ACTIVITY 3-2, page 30

1. Negative
2. Future
3. Negative
4. Future
5. Future
6. Negative
7. Negative
8. Negative
9. Negative
10. Future

UNIT 4

CHECKPOINT 4-1, page 33

1. Past
2. Future
3. Future

4. Past
5. Past

CHECKPOINT 4-2, page 34

1. T
2. F
3. T
4. F
5. T

CHECKPOINT 4-3, page 36

1. I don't want to hear about Kelly.
2. I don't think those jokes are funny. Let's talk about last night's game.
3. Thank you for getting me a can of juice for my break.
4. I appreciate your friendliness and getting an answer for me so quickly.
5. Thank you for calling to tell me the doctor is running late.

CHECKPOINT 4-4, page 38

1. no
2. yes
3. no
4. no
5. yes

CHECKPOINT 4-5, page 40

Your goals may be different than the ones listed. The point is to make the goal SMART.

1. Get a crown on my back molar within 60 days.
2. Make one new friend at work by the next payday.
3. Mow the lawn by Saturday night.
4. Get a college degree by the time I'm 33.
5. Move to a three-bedroom apartment by fall.

ACTIVITY 4-1, page 41

Ask your instructor or a fellow student to read your goals. If they agree that your goals are SMART, give yourself 2 points for each one.

ACTIVITY 4-2, page 41

1. Spend time with positive people; look to the future, not the past; accept yourself; do something you love.
2. You can't change the past. You can learn from the past to make changes in the future.
3. Next time I have a test scheduled, I will start studying earlier so I get a better grade.
4. Notice when he or she does something you like and tell him or her about it.
5. When you thank someone, be specific about what they did. When you set goals, be specific about your goal.
6. You can say, "Thank you," or, "I appreciate. . . . "
7. Don't put others down, don't gossip, don't laugh at sexist or racist jokes, don't take advantage of people with low self-esteem.
8. Specific, Measurable, Achievable, Realistic, Timebound
9. Without a deadline, you may not take action.
10. So you won't forget them. You can review them often.

✔ CHECKING WHAT YOU LEARNED

1. F
2. F
3. T
4. T
5. F
6. T
7. F
8. T
9. F
10. T
11. T
12. T
13. T
14. F
15. T
16. T
17. F
18. F
19. F
20. T
21. T
22. T
23. T
24. T
25. F

PERSONAL PROGRESS RECORD

Name: ______________________________

✔ CHECKING WHAT YOU KNOW

Use the chart below to determine the areas in which you need to do the most work. In the space provided, write the total number of points you got right for each content area. Then add up the total number of points shown in the right column to find your final score. Circle the items you answered correctly. As you begin your study, pay close attention to areas in which you missed half or more of the questions.

Content Area	Item Number	Study Pages	Total Points	Number Right
Self-esteem	1, 2, 3, 5	1-5	8	
Self-image	4, 6	6-7	4	
Self-talk	7	11	2	
Should statements	8, 9, 10	11-12	6	
Overgeneralization	11	13-14	2	
Fortune telling	12	15-16	2	
Catastrophizing	13	17	2	
Fan club	14	21-22	2	
Affirmations-Positive	15, 17	22-23	4	
Visualize Affirmations	16	25-26	2	
Present Tense Affirmations	18, 19	24	4	
When to Use Affirmations	20	27-28	2	
Positive People	21	31	2	
Look to the Future	22	32-33	2	

Content Area	Item Number	Study Pages	Total Points	Number Right
Affirm Others	23	34-35	2	
SMART Goals	24, 25	37	4	

Date ______________________ Total Points: 50 Your Score:

UNIT 1: WHAT IS SELF-ESTEEM?

Exercise	Score
Checkpoint 1-1	________
Checkpoint 1-2	________
Checkpoint 1-3	________
Activity 1-1	________
Activity 1-2	________
TOTAL	________

HOW ARE YOU DOING?

33 or better	Excellent
28-32	Good
23-27	Fair
22 or less	See Instructor

UNIT 2: NEGATIVE THINKING

Exercise	Score
Checkpoint 2 - 1	________
Checkpoint 2 - 2	________
Checkpoint 2 - 3	________
Activity 2 - 1	________
Activity 2 - 2	________
TOTAL	________

HOW ARE YOU DOING?

35 or better	Excellent
30 - 34	Good
25 - 29	Fair
24 or less	See Instructor

UNIT 3: POSITIVE SELF-TALK

Exercise	Score
Checkpoint 3 - 1	________
Checkpoint 3 - 2	________
Checkpoint 3 - 3	________
Checkpoint 3 - 4	________
Activity 3 - 1	________
Activity 3 - 2	________
TOTAL	________

HOW ARE YOU DOING?

38 or more	Excellent
33 - 37	Good
28 - 32	Fair
27 or less	See Instructor

UNIT 4: I LIKE MYSELF!

Exercise	Score
Checkpoint 4 - 1	________
Checkpoint 4 - 2	________
Checkpoint 4 - 3	________
Checkpoint 4 - 4	________
Checkpoint 4 - 5	________
Activity 4 - 1	________
Activity 4 - 2	________
TOTAL	________

HOW ARE YOU DOING?

32 or more	Excellent
29-31	Good
24-28	Fair
23 or less	See Instructor

✔ CHECKING WHAT YOU KNOW

Use the chart below to determine the areas in which you need to do the most review. In the space provided, write the total number of points you got right for each content area. Review those areas in which you missed half or more of the questions. Then add up the total number of points shown in the right column to find your final score.

Content Area	Item Number	Study Pages	Total Points	Number Right
Self-esteem	1, 2, 3, 5	1-5	8	
Self-image	4, 6	6-7	4	
Self-talk	7	11	2	
Should statements	8, 9, 10	11-12	6	
Overgeneralization	11	13-14	2	
Fortune telling	12	15-16	2	
Catastrophizing	13	17	2	
Fan club	14	21-22	2	
Affirmations-Positive	15, 17	22-23	4	
Visualize Affirmations	16	25-26	2	
Present Tense Affirmations	18, 19	24	4	
When to Use Affirmations	20	27-28	2	
Positive People	21	31	2	
Look to the Future	22	32-33	2	
Affirm Others	23	34-35	2	
SMART Goals	24, 25	37	4	

Date ______________________ Total Points: 50 Your Score: ______